Andreas Åkerberg

CFD analyses of the gas flow inside a hot isostatic press

Andreas Åkerberg

CFD analyses of the gas flow inside a hot isostatic press

LAP LAMBERT Academic Publishing

Impressum / Imprint

Bibliografische Information der Deutschen Nationalbibliothek: Die Deutsche Nationalbibliothek verzeichnet diese Publikation in der Deutschen Nationalbibliografie; detaillierte bibliografische Daten sind im Internet über http://dnb.d-nb.de abrufbar.
Alle in diesem Buch genannten Marken und Produktnamen unterliegen warenzeichen-, marken- oder patentrechtlichem Schutz bzw. sind Warenzeichen oder eingetragene Warenzeichen der jeweiligen Inhaber. Die Wiedergabe von Marken, Produktnamen, Gebrauchsnamen, Handelsnamen, Warenbezeichnungen u.s.w. in diesem Werk berechtigt auch ohne besondere Kennzeichnung nicht zu der Annahme, dass solche Namen im Sinne der Warenzeichen- und Markenschutzgesetzgebung als frei zu betrachten wären und daher von jedermann benutzt werden dürften.

Bibliographic information published by the Deutsche Nationalbibliothek: The Deutsche Nationalbibliothek lists this publication in the Deutsche Nationalbibliografie; detailed bibliographic data are available in the Internet at http://dnb.d-nb.de.
Any brand names and product names mentioned in this book are subject to trademark, brand or patent protection and are trademarks or registered trademarks of their respective holders. The use of brand names, product names, common names, trade names, product descriptions etc. even without a particular marking in this works is in no way to be construed to mean that such names may be regarded as unrestricted in respect of trademark and brand protection legislation and could thus be used by anyone.

Coverbild / Cover image: www.ingimage.com

Verlag / Publisher:
LAP LAMBERT Academic Publishing
ist ein Imprint der / is a trademark of
OmniScriptum GmbH & Co. KG
Heinrich-Böcking-Str. 6-8, 66121 Saarbrücken, Deutschland / Germany
Email: info@lap-publishing.com

Herstellung: siehe letzte Seite /
Printed at: see last page
ISBN: 978-3-659-53818-6

Copyright © 2014 OmniScriptum GmbH & Co. KG
Alle Rechte vorbehalten. / All rights reserved. Saarbrücken 2014

ABSTRACT

Hot isostatic pressing (HIP) is a thermal treatment method that is used to consolidate, densify or bond components and materials. Argon gas is commonly used as the pressure medium and is isostatically applied to the material with an excess pressure of 500-2000 bar and a temperature of 500-2200°C. With HIP treatment being a well-established technology for the last decades, one is now striving to obtain an increased understanding of local details in the internal gas flow and heat flux inside the HIP apparatus. The main objective of this work is to assess the potential of using computational fluid dynamics (CFD) as a reliable tool for future HIP development. Two simulations are being performed of which the first one is a steady-state analysis of a phase in the HIP-cycle called sustained state. The second simulation is a transient analysis, aiming to describe the cooling phase in the HIP-cycle. The most suitable modeling approaches are determined through testing and evaluation of methods, models, discretization schemes and other solver parameters. To validate the sustained state simulation, the solution is compared to measurements of operating pressure, heat dissipation rate out through the HIP vessel and local temperature by the vessel wall. However, no validation of the cooling simulations has been conducted. A sensitivity analysis was also performed, from which it could be established that a mesh refinement of strong temperature gradients resulted in an increase of wall heat dissipation rate by 1.8%. Both of the simulation models have shown to yield satisfactory solutions that is consistent with the reality. With the achieved results, CFD has now been introduced into the HIP field and the presented modeling methods may serve as guidelines for future simulations.

TABLE OF CONTENTS

ABSTRACT 1
NOMENCLATURE 4
1 INTRODUCTION 5
   1.1 Objectives and scope of work 5
   1.2 Limitations 6
   1.3 Working method 6
2 OVERVIEW OF THE HIP PROCESS 8
   2.1 Working mechanisms 8
   2.2 Heating phase 9
   2.3 Sustained state 10
   2.4 Cooling phase 10
      2.4.1 Rapid cooling 10
3 THEORY 11
   3.1 Navier-Stokes equations 11
   3.2 Reynolds Averaged Navier-Stokes turbulence modeling 12
      3.2.1 K-ω SST turbulence model 12
   3.3 Near-wall modeling 12
      3.3.1 Dimensionless wall distance $y+$ 13
   3.4 Peng-Robinsons equation of state 14
   3.5 Courant number 14
4 SUSTAINED STATE SIMULATION 15
   4.1 Geometry creation 15
   4.2 Mesh generation 16
   4.3 Case setup 17
      4.3.1 Modeling decisions 18
      4.3.2 Input data 18
      4.3.3 True transient method 21
      4.3.4 Pseudo-transient method 21
      4.3.5 Sensitivity analysis 21
   4.4 Validation 24
5 COOLING PHASE SIMULATION 24
   5.1 Case setup 24
      5.1.1 Input data 24
   5.2 Comparison with measurement data 25
6 RESULTS 25
   6.1 Sustained state results 25

	6.1.1	Method comparison	25
	6.1.2	Pseudo-transient solution	26
	6.1.3	Validation	30
6.2		Cooling phase results	30
	6.2.1	Transient solution	30
7	DISCUSSION AND CONCLUSIONS		33
7.1		Discussion	33
	7.1.1	Sustained state simulation	33
	7.1.2	Cooling phase simulation	33
7.2		Conclusions	34
8	RECOMMENDATIONS AND FUTURE WORK		35
8.1		Recommendations	35
8.2		Future work	35

REFERENCES 37
APPENDIX A: SOLUTION COMPARISON FOR SUSTAINED STATE 38
APPENDIX B: TKE PLOT OF SUSTAINED STATE 40
APPENDIX C: RESULTS FROM SENSITIVITY ANALYSIS 41

NOMENCLATURE

Symbols

ρ	Density	[kg/m³]
t	Time	[s]
u	Velocity component in the x direction	[m/s]
v	Velocity component in the y direction	[m/s]
w	Velocity component in the z direction	[m/s]
μ	Dynamic viscosity	[kg/ms]
p	Static pressure	[Pa]
F	Body force	[N/m³]
c_p	Specific heat at constant pressure	[J/kgK]
T	Temperature	[K]
y	Wall distance	[m]
y^+	Dimensionless wall distance	[-]
u_*	Friction velocity	[m/s]
ν	Local kinematic viscosity	[m²/s]
τ_w	Wall shear stress	[Pa]
σ	Courant number	[-]

Subscripts

x	Component in x direction
y	Component in y direction
z	Component in z direction

Abbreviations

CFD	Computational Fluid Dynamics
EOS	Equation Of State
FSI	Fluid-Structure Interaction
HIP	Hot Isostatic Press/Pressing
N-S	Navier-Stokes
SDR	Specific Dissipation Rate
TKE	Turbulent Kinetic Energy

1 INTRODUCTION

In 1952, ASEA in Sweden manufactured the world's first synthetic diamond out of graphite by employing high temperature and pressure (Zimmerman & Toops, 2008). In order to reach the necessary 70 000 bar (1 015 000 psi) the QUINTUS pressure vessel was invented. It later turned out that the QUINTUS design was an ideal pressure vessel for housing hot isostatic pressing (HIP), which is a thermal treatment method that is used to consolidate, densify or bond components and materials. Argon gas is commonly used as the pressure medium and is isostatically applied to the material with a pressure of 500-2000 bar and a temperature of 500-2200°C.

With HIP treatment being a well-established technology for the last decades, one is now striving to obtain an increased understanding of local details in the internal gas flow and heat flux inside the HIP apparatus. Due to the complex design and the tough operating conditions, which limit the possibilities of implementing measuring equipment, new methods of verifying the flow within the furnace are necessary. In addition, as development pushes the limits for HIP performance where one of the primary focus areas is to reduce the cool-down time, more detailed information is needed for optimal design solutions. This is where computational fluid dynamics (CFD) comes to play an important role in next generation HIP designs.

CFD is a means of modeling and simulating real fluid flows and heat transfer by numerical solving of a set of governing equations. Since the appearance of the first commercial CFD software in the 1980s, the complexity of the applications has increased remarkably as the codes have improved and the available computer power has increased (Peric & Bertram, 2011). CFD-modeling has to the author's knowledge only been used to a limited extent in the field of HIP processing with one similar study being presented in the Proceedings of the 2011 International Conference on Hot Isostatic Pressing (Numerical Simulation of Flow and Thermal Field in a HIP Furnace, 2011). However, with today's improved resources simulations of the internal gas flows and heat fluxes can be performed. If CFD can be shown to be a reliable tool in HIP development, it could aid in acquiring knowledge about the physical phenomena that are taking place inside the apparatus. Additional uses of CFD simulations could also be testing and evaluation of designs alterations and operational parameters without the need for physical modeling.

This work is being conducted for a specific model of a HIP-apparatus that has been manufactured by Avure Technologies, thus descriptions of some of the associated components and functionalities may not be consistent for other HIP models. The studied apparatus involves several phases during one HIP-cycle, of which two are being analyzed in this work; the sustained state and the cooling phase. These phases are explained further in chapter two.

Previous research that has been conducted in the HIP field has mostly been focused to the material structure of the processed components. A small number of similar CFD simulations have however been presented at conferences, which to the author's knowledge only have been steady-state analyses and have not involved any type of transient cooling simulations. Another CFD analysis was found in the literature study, in which a simulation was performed of a heat treatment furnace working at atmospheric pressure and that was heated by oil-burners. (Yang, de Jong, & Reuter, 2005)

1.1 Objectives and scope of work

The main objective of this work is to assess the potential of using CFD as a reliable tool for future HIP development. It is to be investigated if CFD-models can accurately describe the physical phenomena that are taking place inside the HIP apparatus considering gas flow and heat flux. The simulations furthermore have to be computed within reasonable ranges of computational time and effort.

The interior of the HIP pressure vessel is to be simulated in a 2D axisymmetric approach including models for heat and turbulence but excluding a model for radiation. To simplify the problem, the furnace is assumed to contain no load.

Two simulations are to be performed of which the first one is a steady-state analysis of a phase in the HIP-cycle called sustained state. The second simulation is a transient analysis, aiming to describe the cooling phase in the HIP-cycle. The most suitable modeling approaches will be determined through testing and evaluation of methods, models, discretization schemes and other solver parameters.

It should be mentioned that the work is not focused on obtaining highly detailed information on specific parts of the HIP vessel, but to obtain a correct simulation of the overall gas flow and heat flux.

The aim of the work can be summarized in the following objectives:

- Develop modeling approaches for the two simulations that are as time efficient and accurate as possible.
- Obtain a steady state solution describing the sustained state.
- Obtain a transient solution simulating the cooling phase.
- Validate the solutions through comparisons with measurements where possible.
- Assess the potential of using CFD as a reliable tool in future HIP development.

1.2 Limitations

The following factors are assumed to have an impact on the end result of the work:

- Amount of measurement data to validate the CFD model.
- Amount of time and computational power that are available.
- High level of complexity in geometry that needs simplifications.

Validation of the solution is an important part of any CFD analysis as it confirms whether the model can be assumed to be reliable or not. The simulations that are to be performed in this work will be validated by comparing the solutions with measurements of temperature, operating pressure and heat dissipation rate from the HIP-cylinder walls.

The available computer power generally determines the amount of time that is required to compute a simulation. The computational resources and the time available for this work are however considered to be sufficient to yield reliable results within reasonable time.

Simplifications that are considered to have a negligible impact on the final solutions will be performed on the geometry to reduce the mesh size and thereby decrease the required computational time.

1.3 Working method

The working method can be subdivided into the following major steps:

1. Geometry creation
2. Pre-processing of sustained state simulation
3. Solving of sustained state simulation
4. Sensitivity analysis (refine mesh and repeat solving)
5. Pre-processing of cooling phase simulation
6. Solving of cooling phase simulation
7. Post-processing

The same mesh is used for both of the simulations, so the geometry creation and the mesh generation only have to be done once.

The simulations are being created using software from the ANSYS 13.0 package, with the specific applications presented in Table 1.1. The FLUENT solver will furthermore be run in parallel mode utilizing a total of eight cores for the computations.

Table 1.1. Software that is being used in the simulation process.

Software	Task
ANSYS SpaceClaim	Geometry creation
ANSYS Meshing	Mesh generation
ANSYS FLUENT	Pre-processing and solving
ANSYS CFD-Post	Post-processing
ANSYS Workbench	Project management

The modeling process will also involve work segments that do not have a direct impact on the end results but still will be an important part of the developing process. These segments of work will not be discussed further in the report, or only very briefly, and include:

- Study of the software to be used
- Study of the HIP process and equipment
- Study of CFD theory
- Testing and evaluation of meshing techniques
- Testing and evaluation of solving approaches and settings such as schemes, solver types, turbulence models, equations of state, boundary conditions, operating conditions, under-relaxation, solution initialization, time-step and convergence criteria
- Testing and evaluation of post-processing and animation approaches
- Estimations of gas density through extrapolation of available data
- Performing rough hand calculations of certain properties to which the solutions will be compared
- Extraction of measurement data from logged HIP-cycles

The topics are excluded from the report in order to produce a more structured report with a higher relevancy. It should be noted that even though some of the excluded work may not have a significant impact on the end results is still may have taken up a big part of the total project time.

2 OVERVIEW OF THE HIP PROCESS

Some of the purposes of HIP treatment are to

- Eliminate porosity, particularly in castings
- Densify metal and ceramic powders
- Consolidate powder-metallurgy parts

Typical applications include aerospace castings, aluminium castings, body implant castings, armor ceramics, and composite P/M materials. The HIP apparatuses are being manufactured in a wide range of sizes of which one of the bigger constructions is shown in Figure 2.1.

Figure 2.1. HIP apparatus manufactured by Avure Technologies.

2.1 Working mechanisms

The technology enables densification of a powder or preformed material to 100% of its theoretical density (Welding Technology Institute of Australia, 2006) by several working mechanisms. The first mechanism involves the material particles pressing against each other due to the isostatic pressure, so that the particles get plastically deformed. This does however not apply for ceramic materials. In addition to the immediate plastic deformation, time dependent creep is further densifying the material due to glide along crystal planes enhanced by the high temperature. The last porosity is eliminated by the diffusion mechanism that is strongly dependent on temperature. Due to the excess pressure, diffusion may occur at significantly lower temperatures than it commonly does. (ASEA Metallurgy, 1984)

A simplification of the interior of the HIP vessel is shown in Figure 2.2. The solid parts in the equipment consist of the steel alloy SA-273, except for where a high-temperature-resistant material is necessary, where instead graphite or molybdenum is used. The insulation mantle has solid exterior surfaces and a porous insulation interior. Gas is allowed to go into the mantle and move around in the porous insulation material, which heavily affects the thermal properties of the insulation mantle.

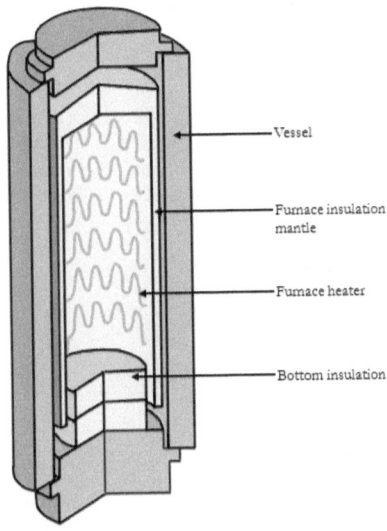

Figure 2.2. Pressure vessel of a HIP apparatus without fan-enhanced or ejector-enhanced cooling system

The load that is to be treated is placed in the furnace that is enclosed by the insulation mantle. The load can have a specific loading structure, but can also be placed in a randomized manner. For some types of load, it is important that the gas has a uniform temperature inside the furnace so that all parts of the load will be equally treated. The manufacturer ensures that the maximum temperature difference in the loading section should not exceed 8°C in the specific HIP-apparatus that is being investigated.

A typical HIP-cycle is presented in Figure 2.3 below. It should be mentioned that the heating phase usually involves simultaneous pumping of gas as well. However, splitting of the heating and pumping facilitates calculations.

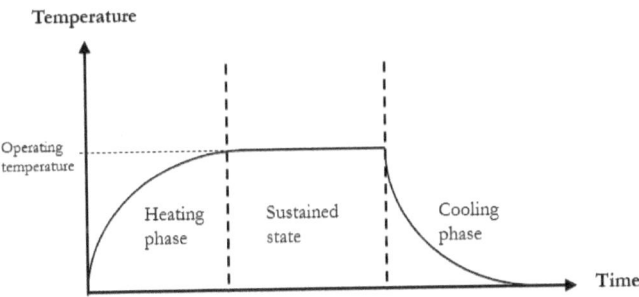

Figure 2.3. The three main phases during the HIP-cycle.

2.2 Heating phase

As the heating phase begins, the HIP vessel has already been filled with the pressure medium, which consists of an inert gas with an initial excess pressure. The pressurized gas is heated by electrical heaters, which consist of current-carrying wires. The heating phase ends when the desired operating temperature and operating pressure have been reached.

2.3 Sustained state

The HIP-cycle enters sustained state when the furnace has been fully heated to the operating temperature and operating pressure. As the densifying process is time-dependent, the operating temperature and pressure have to be held sustained for a specific amount of time until the desired material properties have been obtained. The furnace is constantly being heated to retain the operating temperature and the vessel is being chilled from the outside to protect it from overheating.

2.4 Cooling phase

When the processed components have been exposed to heat and pressure for a sufficient amount of time, the HIP is being cooled down so that the furnace can be emptied on gas and the processed components can be unloaded. The cooling of the furnace has traditionally been performed naturally without any assisting cooling equipment inside the vessel. This method is however a time consuming approach which gives a low productivity.

2.4.1 Rapid cooling

A higher cooling rate can be achieved during the cooling phase by using a fan or a compressor to quickly transport gas from a cold zone in the vessel, to the hot furnace zone. A high cooling rate is of high interest since it implicates that more HIP-cycles can be run in the same amount of time, and thereby the productivity can be increased for the users.

3 THEORY

This chapter presents the general underlying theory for CFD and other subtopics that are relevant for this work.

3.1 Navier-Stokes equations

CFD provides a way to model and simulate real fluid flows and heat transfer by numerical solving of a set of governing equations. The fluid space that is to be modeled, known as the computational domain, is discretized into a grid of cells that is called mesh. The core of CFD is the Navier-Stokes (N-S) equations which describe the motion of fluids (Sayma, 2009) and are solved for each cell in the mesh. The N-S equations primarily consist of the conservation equations for mass and momentum, which are solved for all flow problems in FLUENT.

Conservation of mass:

$$-\frac{\partial \rho}{\partial t} = \frac{\partial(\rho u)}{\partial x} + \frac{\partial(\rho v)}{\partial y} + \frac{\partial(\rho w)}{\partial z} \quad (1)$$

Conservation of momentum:

$$\rho\left(\frac{\partial u}{\partial t} + u\frac{\partial u}{\partial x} + v\frac{\partial u}{\partial y} + w\frac{\partial u}{\partial z}\right) = -\frac{\partial p}{\partial x} + \mu\left(\frac{\partial^2 u}{\partial x^2} + \frac{\partial^2 u}{\partial y^2} + \frac{\partial^2 u}{\partial z^2}\right) + F_x \quad (2)$$

$$\rho\left(\frac{\partial v}{\partial t} + u\frac{\partial v}{\partial x} + v\frac{\partial v}{\partial y} + w\frac{\partial v}{\partial z}\right) = -\frac{\partial p}{\partial y} + \mu\left(\frac{\partial^2 v}{\partial x^2} + \frac{\partial^2 v}{\partial y^2} + \frac{\partial^2 v}{\partial z^2}\right) + F_y \quad (3)$$

$$\rho\left(\frac{\partial w}{\partial t} + u\frac{\partial w}{\partial x} + v\frac{\partial w}{\partial y} + w\frac{\partial w}{\partial z}\right) = -\frac{\partial p}{\partial z} + \mu\left(\frac{\partial^2 w}{\partial x^2} + \frac{\partial^2 w}{\partial y^2} + \frac{\partial^2 w}{\partial z^2}\right) + F_z \quad (4)$$

In these equations u, v and w are the velocity components in the x, y and z directions. ρ is the density, p is the pressure and μ is the dynamic viscosity.

If heat is involved in the problem, as it is in this work, an additional energy equation is solved as well.

Conservation of energy:

$$\rho c_p \left(\frac{\partial T}{\partial t} + u\frac{\partial T}{\partial x} + v\frac{\partial T}{\partial y} + w\frac{\partial T}{\partial z}\right)$$
$$= \Phi + \frac{\partial}{\partial x}\left[k\frac{\partial T}{\partial x}\right] + \frac{\partial}{\partial y}\left[k\frac{\partial T}{\partial y}\right] + \frac{\partial}{\partial z}\left[k\frac{\partial T}{\partial z}\right] + \left(u\frac{\partial p}{\partial x} + v\frac{\partial p}{\partial y} + w\frac{\partial p}{\partial z}\right) \quad (5)$$

where c_p is the specific heat at constant pressure, T is the temperature and Φ is the dissipation function given by

$$\Phi = 2\mu\left[\left(\frac{\partial u}{\partial x}\right)^2 + \left(\frac{\partial v}{\partial y}\right)^2 + \left(\frac{\partial w}{\partial z}\right)^2 + 0.5\left(\frac{\partial u}{\partial y} + \frac{\partial v}{\partial x}\right)^2 + 0.5\left(\frac{\partial v}{\partial z} + \frac{\partial w}{\partial y}\right)^2 \right.$$
$$\left. + 0.5\left(\frac{\partial w}{\partial x} + \frac{\partial u}{\partial z}\right)^2\right] - \frac{2}{3}\mu\left(\frac{\partial u}{\partial x} + \frac{\partial v}{\partial y} + \frac{\partial w}{\partial z}\right)^2 \quad (6)$$

As the N-S equations are discretized and solved for each cell, flow variables, such as velocity and temperature, will be obtained at the discrete locations.

3.2 Reynolds Averaged Navier-Stokes turbulence modeling

To effectively model turbulent flows, some alterations are usually being made to the N-S equations so that they become the Reynolds Averaged Navier-Stokes (RANS) equations, which instead yield approximate time-averaged solutions. The principle of Reynolds averaging is to decompose every scalar quantity in the instantaneous N-S equations into one time-averaged mean component and one fluctuating component in the following way

$$\phi = \bar{\phi} + \phi' \tag{7}$$

ϕ is an arbitrary scalar such as pressure, energy or species concentration. $\bar{\phi}$ is the mean component and ϕ' is the fluctuating component.

The velocity scalar gets for instance decomposed as

$$u_i = \bar{u}_i + u'_i \quad i = 1,2,3 \tag{8}$$

By substituting these decomposed flow variables into the N-S equations and time-averaging the equations, the resulting RANS equations in Cartesian tensor form become as shown below.

Convervation of mass:

$$\frac{\partial \rho}{\partial t} + \frac{\partial}{\partial x_i}(\rho u_i) = 0 \tag{9}$$

Conservation of momentum:

$$\frac{\partial}{\partial t}(\rho u_i) + \frac{\partial}{\partial x_j}(\rho u_i u_j)$$
$$= -\frac{\partial p}{\partial x_i} + \frac{\partial}{\partial x_j}\left[\mu\left(\frac{\partial u_i}{\partial x_j} + \frac{\partial u_j}{\partial x_i} - \frac{2}{3}\delta_{ij}\frac{\partial u_l}{\partial x_l}\right)\right] + \frac{\partial}{\partial x_j}(-\rho \overline{u'_i u'_j}) \tag{10}$$

Compared to the regular N-S equations an additional term has now appeared, $-\rho \overline{u'_i u'_j}$, which is called the Reynolds stresses and has to be modeled by a turbulence model.

Two alternative methods to RANS modeling are large eddy simulation (LES) and direct numerical simulation (DNS), though these will not be considered in this work.

3.2.1 K-ω SST turbulence model

The shear stress transport (SST) k-ω model combines the standard k-ω model and the k-ε model as the two formulations complement each other very well. The standard k-ω formulation efficiently describes the near-wall regions but is very sensitive to the turbulence properties in the free-stream which instead the k-ε formulation models very accurately. The SST k-ω model therefore switches to a k-ε formulation in the free-stream as it performs better in those regions.

The SST k-ω model is a two-equation model, meaning that it adds two additional transport equations to the problem to describe the turbulent properties in the flow. The added transported variables are the turbulent kinetic energy, k, and the specific dissipation, ω. (ANSYS Inc.)

3.3 Near-wall modeling

Near-wall modeling is an important part of the modeling as an incorrectly modeled boundary layer can have an impact on the overall fluid flow and result in inaccurate results. The fluid region close to the wall can be subdivided into two important sub-layers as shown in Figure 3.1 (ANSYS Inc.). The innermost layer is called the viscous sub-layer in which the flow is almost laminar and the viscosity is dominant in transfer of momentum and heat. Outside the sub-layer follows the logarithmic layer, in which the mixing process is instead dominated by turbulence. An additional region should also be mentioned, which is the buffer layer that is located between the sub-layer and the logarithmic layer in which both viscous and turbulent effects are equally important.

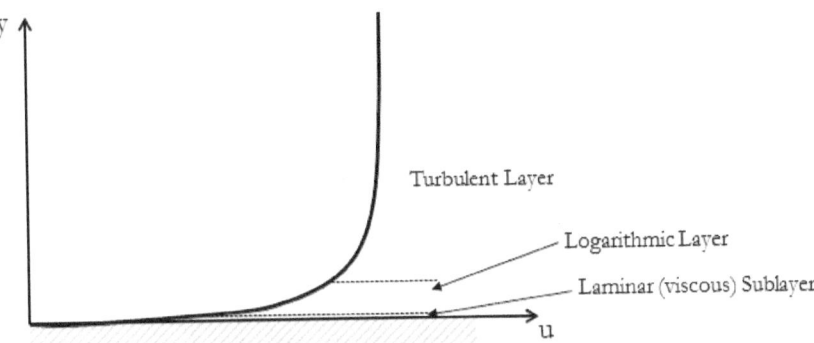

Figure 3.1. Subdivision of the near-wall region. Distance from the wall, y, on the vertical axis and velocity, u, on the horizontal axis.

The region can be modeled either by resolving it using a very fine grid near the wall or by using semi-empirical formulas called wall functions which will describe the viscous sub layer through the law of the wall (Rao, 2010). The ANSYS documentation highly recommends however that at least three nodes should be used to resolve the boundary layer.

3.3.1 Dimensionless wall distance y+

As mentioned, wall functions will have to be used to describe the boundary layer if the mesh is not fine enough close to the walls. ANSYS FLUENT determines whether wall functions will be used or not depending on the distance between the wall and the location of the first node in the mesh. This distance is expressed by the dimensionless wall distance y^+, which is defined as

$$y^+ \equiv \frac{u_* y}{v} \qquad (11)$$

y is the distance to the nearest wall, v is the local kinematic viscosity and u_* is the friction velocity, which is defined as

$$u_* \equiv \sqrt{\frac{\tau_w}{\rho}} \qquad (12)$$

where τ_w is the wall shear stress and ρ is the fluid density at the wall.

An y+ value less than or equal to 11 implicates that the first grid node is located within the laminar sub-layer and a wall function will thereby not be used. For values greater than 11, the region between the wall and the first cell node is assumed to have a velocity profile with a logarithmic shape, i.e. being modeled with a wall function.

3.4 Peng-Robinsons equation of state

An equation of state (EOS) is used to calculate the fluid density depending on temperature and pressure. The simplest and most common EOS is the ideal gas law. The ideal gas law is however not applicable when the pressure exceeds the critical pressure of the fluid (ANSYS Inc.). This is unfortunately the case for the simulations in this work since the critical pressure for argon is 49 bar and the operating pressure is over 1000 bars.

A more complex EOS-alternative is the Peng-Robinson real gas model, which applies also for higher temperatures and pressures, compared to the ideal gas law. Comparisons with argon density data have shown that the Peng-Robinson equation gives reliable density values in the simulated conditions.

3.5 Courant number

The Courant number is a term that appears when discretizing transport equations in space with explicit advancing in time from time step n to $n+1$ (Sayma, 2009). The Courant number is generally defined as

$$\sigma = \frac{u \Delta t}{\Delta x} \tag{13}$$

Where u is velocity or convection speed, Δt is time step and Δx is grid size.

A high Courant number implicates a faster advancement in time with risk for instabilities, which can lead to increasing errors with time. A lower number implicates a slower advancement in time requiring more computational time but reduces instabilities.

4 SUSTAINED STATE SIMULATION

The first simulation is a steady state analysis of the sustained state in which the vessel has been fully heated to the desired temperature and pressure. The actual heating phase that precedes the sustained state in the HIP-cycle is not considered in this work.

4.1 Geometry creation

The geometry is created within the ANSYS SpaceClaim software and is based on a drawing of the analyzed HIP apparatus. Scaling of the drawing and simplifications of the geometry are applied as necessary. The vessel is being modeled in a 2D axisymmetric approach, which is considered to be valid due to the almost completely axisymmetric geometry of the real vessel. Using this simplification it is only necessary to create the geometry for half of the cross-sectional plane across the HIP as shown in Figure 4.1.

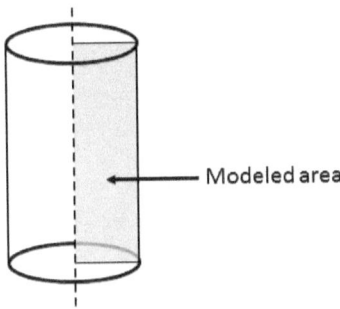

Figure 4.1. Geometry structure in an axisymmetric model.

Some geometry details are completely removed as they are only appearing sporadic around the symmetry axis and therefore are considered to have a negligible impact on the overall flow.

Other geometry details may not be axisymmetric but follow a periodic pattern around the symmetry axis. In these cases the periodic details are substituted with an equivalent continuous geometry around the axis, as illustrated in Figure 4.2.

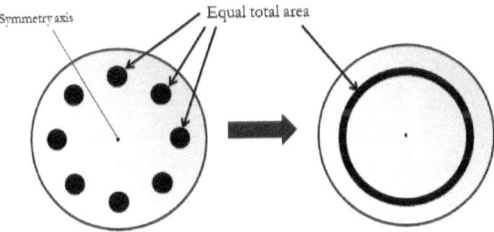

Figure 4.2. A geometry which follows a periodic pattern around the symmetry axis that being simplified as a continuous shape with an equivalent total area.

The geometry is further simplified by removing smaller cavities in the solids that are considered to have a negligible impact on the flow while forcing the mesh into smaller cell sizes and thereby increasing the required computational time.

As the used version of ANSYS Meshing does not have the possibility to create new zones the zones have to be created directly in the geometry. The fluid part of the geometry is split up into several zones to give more control in the meshing process. Some of the solid regions are split up into several zones as well to be able to apply different materials to different parts of the solid.

The geometry is oriented so that the symmetry axis is aligned with the x-axis in the global coordinate system which is a requirement for axisymmetric simulations in FLUENT. (ANSYS Inc.)

4.2 Mesh generation

The mesh generation process only has to be performed once since the same mesh is used for simulating both the sustained state and the cooling phase. The geometry is imported into the ANSYS Meshing software. The aim of the meshing process is to produce a mesh that may be used to compute the correct solution with a minimum number of cells.

The procedure follows the following steps:

1. Generate structured mesh of the fluid in the channel zones
2. Generate unstructured mesh for the remaining fluid
3. Generate structured and unstructured mesh for the solids

As correct capturing of the channel flow behaviors is of high importance for the end result, the structured mesh in the channel zones are generated first. The surrounding mesh will therefore have to adapt to the cells that have already been generated in the channels.

The mesh in the solid areas is preferably generated last as the solids will only transport heat and therefore do not need to have a very smooth or fine mesh and therefor can adapt to the already generated fluid mesh.

A part of the final mesh, showing the half side of the furnace, can be seen in Figure 4.3. It has been found effective to use a structured mesh in all channels since the direction of the flow is known there and the cells can therefore be aligned with the flow. An unstructured mesh is used in the remaining fluid zones due to the complexity of the geometry and the inability to predict the flow. This minimizes the amount the numerical diffusion in the channels as diagonally aligned cells will be avoided.

Triangles are in particular used for large solid parts as they enable a smooth transition from the high mesh resolution in the fluid zones to a coarser resolution, which is sufficient inside the solid zones. Triangles however also occur occasionally in mixed mesh-zones where both quadrilaterals and triangles can be used.

Mesh statistics are presented in Table 4.1.

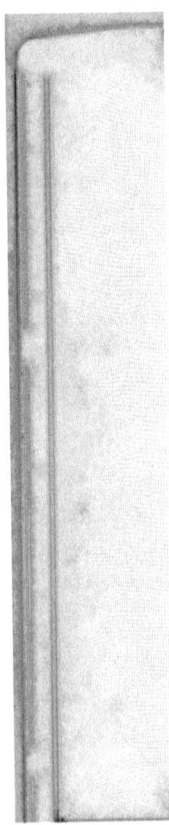

Figure 4.3. Part of the final mesh showing the half side of the furnace.

Table 4.1. Mesh statistics

Parameter	Value		
Element types	Quadrilaterals and triangles		
Nodes	280016		
Elements	294943		
	Min	Max	Average
Orthogonal quality	0,32	1	0,98
Skewness	1,31e-10	0,83	9,02e-02
Aspect ratio	1	11,353	1,96

4.3 Case setup

Since the sustained state is in thermodynamic equilibrium, it implicates that the desired solution can be classified as steady state. However, due to the complexity of the case, a regular steady state simulation is difficult to perform, especially without an initial solution that is very close to the final solution. Instead of a steady state-solving approach, two different transient approaches are instead being used to solve for the steady state solution.

The first method, the true transient approach, is the most accurate method while the second one, the pseudo-transient approach, is considered to be more time efficient but is not necessarily as reliable as the first method.

Even though the system is in steady state, a true transient approach is used to step forward in time to the final steady state. The reason why it is used is because it will take important phenomena into account that may occur during the heating process of the furnace, which could affect the final solution. A regular steady state-solving approach, which solves for the final state directly, would not take these kinds of phenomena into account.

The pseudo-transient approach is a hybrid between regular steady state solving and transient solving and it has the ability to calculate with different time steps for the fluid zones and the solid zones.

The solutions of the two methods will be compared to determine if the pseudo-transient approach is able to model the case with the same accuracy as the true transient method.

4.3.1 Modeling decisions

As the heaters consist of permeable wires they cannot be modeled as impermeable solid zones. The heaters will therefore be defined as fluid zones and the heater boundaries will be defined as interior, which will allow the gas move in and out of the heater zone.

The insulation mantle consists of an enclosing solid surface with porous insulation material inside. Gas is allowed to go into the mantle and move around in the porous material, which will heavily affect the thermal properties of the mantle. This will however not be considered in the model and instead an experience-based value of the thermal conductivity is used for the whole mantle.

4.3.2 Input data

The boundary conditions are visualized in Figure 4.4.

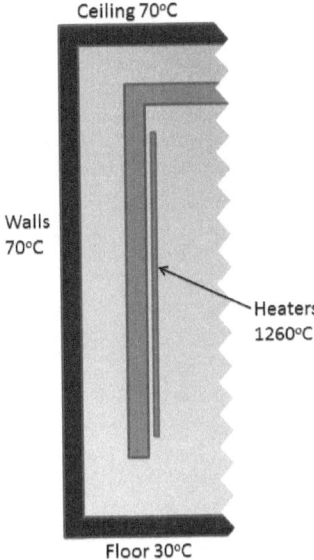

Figure 4.4. Boundary conditions for the sustained state simulation.

The boundary conditions introduce some further simplifications to the model. In reality, the heaters have a decreasing temperature towards the bottom of the vessel due to buoyancy effects, while in the model they are simplified to have a constant temperature from top to bottom. The surrounding walls, the floor and the ceiling have also been simplified to have constant temperatures, which is inconsistent with reality as well.

All of the outer boundaries of the computational domain are of the boundary type wall, except for the symmetry axis which is set to boundary type axis. All the walls have the no slip-condition enabled which implicates that the fluid at the solid boundary has zero velocity relative to the wall.

General input data for both the true transient and pseudo-transient model is presented in Table 4.2.

Table 4.2. Input data that is shared between the two simulation approaches for the sustained state.

Parameter	Value
Near-wall modeling	Log-law wall function
k-ω SST model constants	Standard
Operating pressure	Floating or Fixed
Pressure-velocity coupling	Coupled algorithm
Gradient discretization scheme	Least squares cell based
Pressure discretization scheme	PRESTO!
Density discretization scheme	2nd order
Momentum discretization scheme	2nd order
TKE discretization scheme	2nd order
SDR discretization scheme	2nd order
Energy discretization scheme	2nd order

Wall functions are chosen to model near-wall regions to reduce the mesh size. Furthermore the k-ω SST turbulence model is chosen due to its versatility in both near-wall regions and free-streams. The operating pressure is set to floating for the true transient approach so that it can rise with the increasing temperature. The pressure does not need to be floating in the pseudo-transient simulation but can instead be set to a fixed value obtained from the true transient solution. The pressure-velocity coupling is set to a coupled algorithm as it appear to give an increased stability and provides the possibility to set a Courant number which can be altered to further adjust stability and convergence speed.

The least squares cell based algorithm for gradient discretization is the default option in FLUENT. The PRESTO!-scheme is chosen for pressure discretization which is recommended for high-Rayleigh-number natural convection problems (ANSYS Inc.). The remaining discretized variables are density, momentum, turbulent kinetic energy (TKE), specific dissipation rate (SDR) and energy which are set to use schemes of second order to yield a higher accuracy of the solution and to eliminate numerical diffusion (Bakker, 2002). As this case includes very strong gradients of temperature, pressure and density it is very important to minimize the occurrence of numerical diffusion.

Some major differences in the setups of the true transient case and the pseudo-transient case can be identified. These differences are presented in the table below.

Table 4.3. Main differences in solving methods between the true transient approach and the pseudo-transient approach.

	True transient	Pseudo-transient
Initial solution	400 bar, 50°C	1311 bar, 1260°C
Operating pressure	Floating	Fixed at 1311 bar
Solution type	Transient but unchanged with further time steps	Steady state

The thermal physical properties of the materials are presented in Table 4.4 below.

Table 4.4. Thermal physical properties of gas and solid materials.

Property	Value
Argon density	Derived from Peng-Robinsons equation
Argon specific heat	520.64 J/kgK
Argon thermal conductivity	0.073 W/mK
Argon viscosity	2.125e-5 kg/ms
Additional argon properties	FLUENT database standard
SA-723 density	7750 kg/m^3
SA-723 specific heat	502.48 J/kgK
SA-723 thermal conductivity	40 W/mK
Isolation density	128 m^3
Isolation specific heat	830 J/kgK
Isolation thermal conductivity	0.5 W/mK
Molybdenum density	10188 kg/m^3
Molybdenum specific heat	250 J/kgK
Molybdenum thermal conductivity	138 W/mK

The argon density will be computed continuously in the simulation from the EOS, which is the Peng-Robinsons equation. This EOS has been chosen since it has been verified to apply for high temperatures and pressures, conditions at which for instance the ideal gas law is not valid. The computed densities are being verified through comparisons with extrapolated measurement data that has been provided by Avure Technologies.

4.3.3 True transient method

The true transient method involves a fictional heating phase in the simulation, which resembles the real heating phase described in chapter 2. The primary difference between the fictional heating phase and the real heating phase is that the simulated heaters have their temperatures fixed at the end temperatures already from the beginning of the phase, while the real heaters gradually increase in temperature.

The solution is initialized to a pressure of 402 bar and a temperature of 50°C in the complete computational domain. As the calculation begins, the furnace is then heated gradually until steady state conditions have been achieved.

The case is set up to have a floating operating pressure, which enables the operating pressure to gradually increase as the system gets heated through calculations of the integral mass balance (ANSYS Inc.).

In the beginning of the simulation, the time step is chosen and adjusted continuously to result in a maximum cell courant number close to 1. When the flow seems to have stabilized and the simulation is primarily waiting for the heat conduction in the solids to finish, the courant number is increased gradually from 1 to 10. An additional criterion when choosing the time step is also that the smallest residual decrease of all the solved variables should have a residual decrease of two to three orders of magnitude.

To determine that convergence has been achieved in a transient simulation, a widely used approach is to setup monitors of important flow properties such as fluid velocity or heat transfer rate. When the monitored properties have stabilized and do not seem to change with further time steps, the solution can be assumed to have converged. The flow properties that are monitored in the transient solving of the sustained state are

- Heat dissipation rate out through the vessel
- Floating operating pressure
- Flow velocity in cooling channels

4.3.4 Pseudo-transient method

Unlike to the true transient approach, this method requires an initial solution that is supposed to be a guess of the final solution that should be as good as possible. Therefore, the final floating pressure, which has been calculated in the previous solution, is now used for the initial pressure as well as for a fixed operating pressure. The initial temperature is set to 1260°C in the whole computational domain. To improve the initial guess and thereby decreasing the necessary computational time for the calculation, the energy equation, and the energy equation only, is first solved separately using a regular steady state approach. This solves the temperature field over the computational domain close to the state of the final solution.

4.3.5 Sensitivity analysis

One of the main problems with CFD analyses in general is to prove that the solution is accurate. Many sources of potential error exist that can affect the accuracy of the solution. One of them is bad meshing, which if the mesh is of too poor quality, will give a faulty solution even though the rest of the case is set up correctly. To make sure that the mesh is of sufficient resolution and quality to simulate the problem, a sensitivity analysis of the mesh is to be performed after a solution for the sustained state has been obtained. The sensitivity analysis is being performed by further refining the mesh in key locations and then solve for the solution again. If the solution remains the same, it implicates that the solution is independent of the performed refinements and that they are therefore not affecting the accuracy of the solution. Optimally, a full refinement of the complete mesh should be performed, but is not a possibility in this work due to time constraints.

A sensitivity analysis of the mesh is being conducted based on the pseudo-transient solution by using the following adaption features in the FLUENT software:

- Refinement of strong temperature gradients
- Smoother cell size changes
- Overall refinement of big fluid areas

The impacts of the different adaption methods are presented in the results chapter.

4.3.5.1 Sensitivity analysis results

Several refinement types were performed and evaluated. Only one of them did however have a significant impact on the solution and that was the refinement of strong temperature gradients. Through experimentation and evaluation the following four-step adaption approach were developed and shown to give an increased accuracy in the solution in important areas while also giving a decrease in continuity and energy residuals of more than one order of magnitude.

1. **Gradient adaption**

   The input data to perform the adaption is presented in Table 4.5 below. The refine threshold was adjusted to capture most of the strong temperature gradients and equals approximately one hundredth of the maximum temperature gradient value, which can be computed by the adaption function of FLUENT.

Table 4.5. Input data for the gradient adaption function in FLUENT.

Parameter	Value
Gradients of	Static temperature
Refine threshold	0.002
Zones marked for adaption	Only fluid zones
Other input data	Standard settings

2. **Volume adaption**

   Configure to mark cells based on volume change with the standard maximum volume change of 2.5. The volume adaption is also performed only in fluid zones.
3. **Repeat gradient adaption**

   Same configurations as the previous gradient adaption.
4. **Repeat volume adaption**

   Same configurations as the previous volume adaption.

The resulting mesh after each adaption step is shown in the image sequence in Figure 4.5.

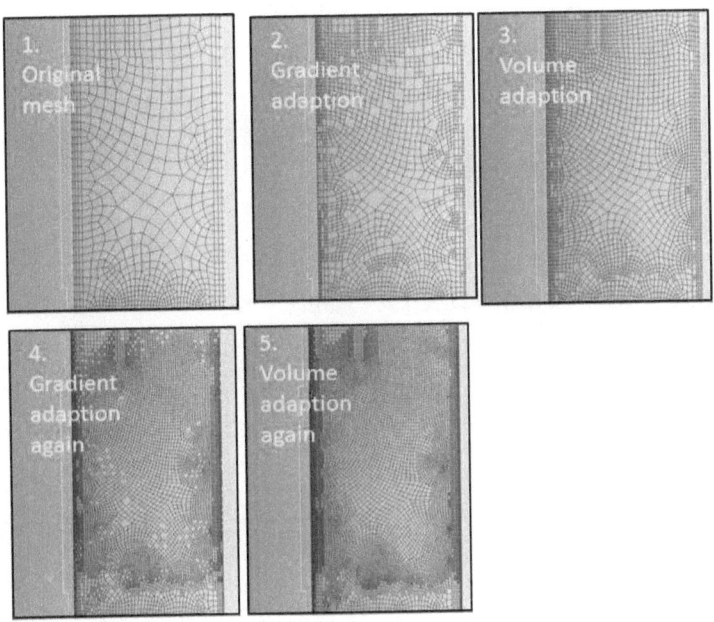

Figure 4.5. The mesh impact of the gradient adaption and the volume adaption.

The mesh refinement resulted in an increase of the total heat dissipation rate as presented in Table 4.6. More refinements than two did not have any further impact.

Table 4.6. The impact of the mesh refinements on the total heat dissipation rate out through the vessel walls.

	Before mesh refinement	After two mesh refinements	After an additional mesh refinement
Total heat dissipation rate	45431 W	46263 W	46238 W

Figures showing the differences in the solutions before and after the mesh refinement are presented in

APPENDIX C: RESULTS FROM SENSITIVITY ANALYSIS.

4.4 Validation

To prove the reliability of the performed simulations, validations of the solutions will be performed through comparisons with measurement data. The simulations will be validated by comparing

- Heat dissipation rate out through the vessel
- Operating pressure inside the vessel
- Local temperature at the vessel wall

5 COOLING PHASE SIMULATION

5.1 Case setup

Since both the sustained state simulation and the cooling phase simulation share the same computational domain and geometry, the solution obtained from the sustained state simulation can therefore be imported directly into the cooling phase case and be used as an initial solution.

After the sustained state simulations had been performed, it was however discovered that some alterations to the geometry were necessary. The alterations comprised a more simplified ejector design and the small gas volume to which the ejector flow will be applied.

5.1.1 Input data

The cooling phase simulation is aiming to simulate the process from the point where the vessel is fully heated and the ejector is started, to the fully cooled state. The converged solution from the previous finished simulation is used as an initial solution in this case setup. The same case setup is used with some minor changes which are presented in the table below.

Table 5.1. Solver options that have changed in the cooling phase simulation since the previous pseudo-transient simulation of the sustained state.

Option	New setting
Time	Transient
Fixed temperatures at heaters	Removed
Operating pressure	Floating
Pressure-velocity coupling	SIMPLEC
Ejector velocity	Fixed velocity applied

The solver is now set to perform a transient analysis instead of a steady state analysis. As the cooling phase begins, the heaters are turned off so the fixed heater temperatures therefore have to be removed. As the pressure has to be allowed to decrease with the decreasing temperature, the operating pressure is set to floating. The pressure-velocity coupling is changed from Coupled to SIMPLEC, which is a segregated algorithm. A segregated algorithm is necessary to be able to apply a fixed velocity to a fluid zone (ANSYS Inc.). The cooling process is working by an induced cooling flow from the ejector, which is simulated by applying a fixed velocity to a fluid zone in the middle of the ejector channel. To decrease the instantaneous shock in the system due to the sudden velocity leap, the ejector velocity is ramped up to the final velocity of approximately 4 m/s in 25 time steps of about 0.0001 s.

The remaining solver settings are left unchanged since the sustained state simulation, including the material properties and the remaining boundary conditions.

The time step and the amount of iterations per time step are being adjusted continuously to fulfill the following criteria:

- The smallest residual decrease of all the solved variables should have a residual decrease of two to three orders of magnitude.
- The maximum cell Courant number in the mesh should be close to 1 in the beginning of the simulation and is gradually allowed to increase to approximately 10.

The time step generally lies in the range of 0.5-1.5 ms for the most part of the time.

5.2 Comparison with measurement data

The resulting outcome of the cooling simulation is highly dependent on the total thermal mass of the system. The thermal mass is in turn affecting to a high degree by the amount of load that is placed in the furnace. While the performed simulation is not considering any form of load in the furnace, the measurements that are available are all of them having some form of load during the cycles.

Unfortunately, this leads to difficulties for the validation part of the cooling simulation and only rough comparisons are therefore to be performed.

6 RESULTS

This chapter presents the solutions from the two performed simulations. Results from the method comparison for the sustained state simulation are also being presented briefly. To avoid revealing of confidential information, which is essential that it is kept only within the company, some parts of the results can unfortunately not be presented in this report. The presented property plots are therefore being delimited to only enclose the absolute hottest part of the furnace including the loading section and the heaters. Normalized properties have furthermore been used where necessary.

6.1 Sustained state results

The sustained state was solved using two different methods, which solutions are compared in the following paragraph. The final solution for the sustained state simulation is then being presented.

6.1.1 Method comparison

The two different methods that were used to solve for the sustained state were the true transient approach, which had a computational time of approximately 3 weeks, and the pseudo-transient approach, which required approximately 24 hours. If the two methods are shown to yield close to identical solutions, the pseudo-transient approach can thereby be considered to be as reliable as the true transient method. The pseudo-transient method may then be used in future simulations to save a significant amount of time compared to if the true transient method is being used.

The following table briefly presents the differences between the solutions in the regards of temperature, density, velocity and turbulent kinetic energy. The overall difference applies in almost the entire computational domain while the maximum difference is highly local and only applies in a few cells. The complete graphical comparisons from which the numbers are extracted can be found in APPENDIX A: SOLUTION COMPARISON.

Table 6.1. The resulting differences after subtracting the pseudo-transient solution from the true transient solution.

Flow variable	Overall difference	Maximum difference
Temperature	< 3 K	40 K
Density	< 3 kg/m³	20 kg/m³

Velocity	~0 m/s	0.2 m/s
Turbulent Kinetic energy	< 0.001 J/kg	0.0018 J/kg

Considering the percentual differences of all the compared properties, they are generally less than 1% of the maximum value of that property.

6.1.2 Pseudo-transient solution

As the pseudo-transient solution showed minor differences to the true transient solution, any of the two solutions can be used to present the results. The pseudo-transient solution was however chosen as the source for the post-processing work.

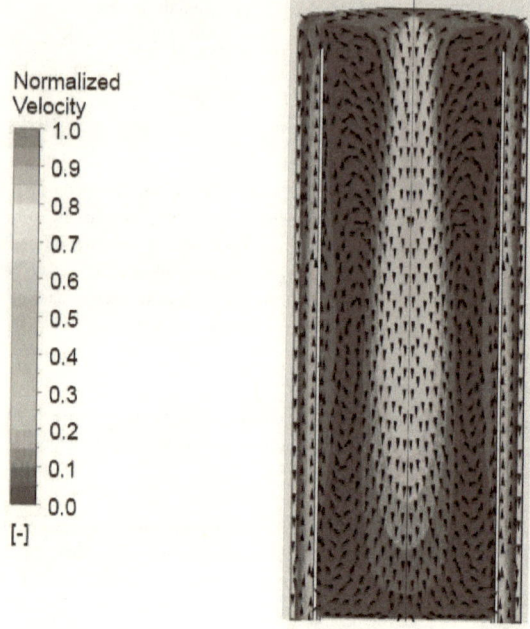

Figure 6.1. Contour of velocity magnitude in the furnace overlaid with black normalized vectors showing velocity direction.

An overview of the overall gas flow in the whole furnace can be seen in Figure 6.1. The two separable loops, the inner loop and the outer loop, can clearly be distinguished. The inner loop rises when the gas is heated by the heaters and falls down in the load section where a significant gas transport can be seen. In the outer loop the gas rises in the inner cooling channel and falls in the outer cooling channel.

While some large vortices can be distinguished in the velocity field, the resolution of this overall plot is too coarse to identify smaller eddies.

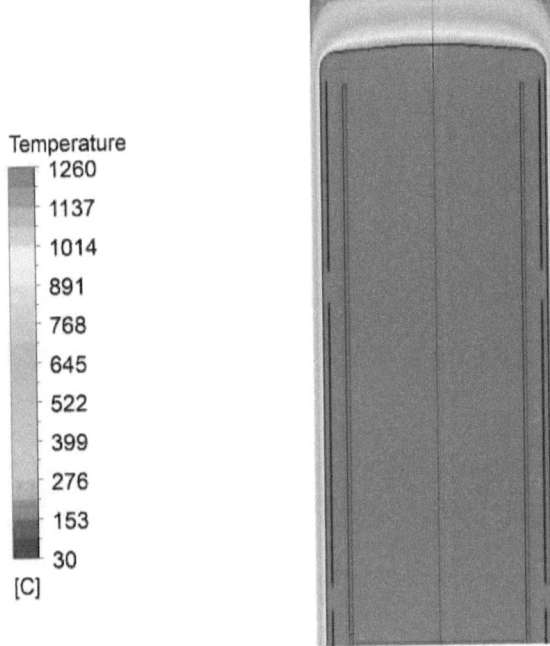

Figure 6.2. Temperature contour of the gas and solids in the furnace during sustained state.

In Figure 6.2, two general temperature zones can be identified, a hot zone and a cold zone. These are completely separated by insulation except in two specific areas, which are characterized by their distinct temperature gradients in the gas. These temperature gradients arise due to the gas in these areas have a velocity close to zero, which is also verified by velocity plot in Figure 6.1. The close-to-zero velocity of the gas results in the heat transfer only taking the form of conduction without any influence of convection. The absence of convection also implicates that no significant amount of hot gas may slip through from the hot zone to the cold zone.

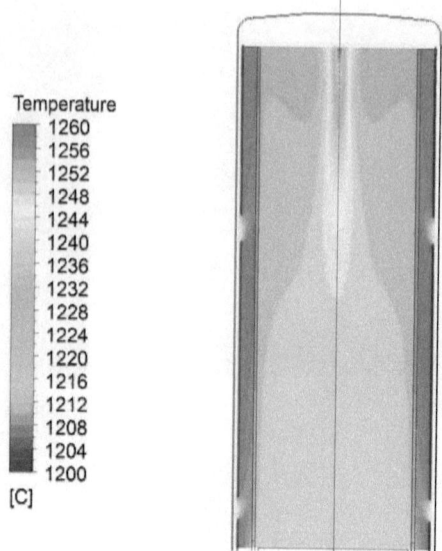

Figure 6.3. Temperature contour of only the hottest area of the vessel during sustained state.

A better resolution of the hottest zone in the furnace is shown in Figure 6.3, including the heaters and the loading section. As expected, the highest temperatures are obtained right by the heaters and slightly colder temperatures are appearing in the loading area. A rather significant temperature decrease is furthermore appearing at the top center of the furnace.

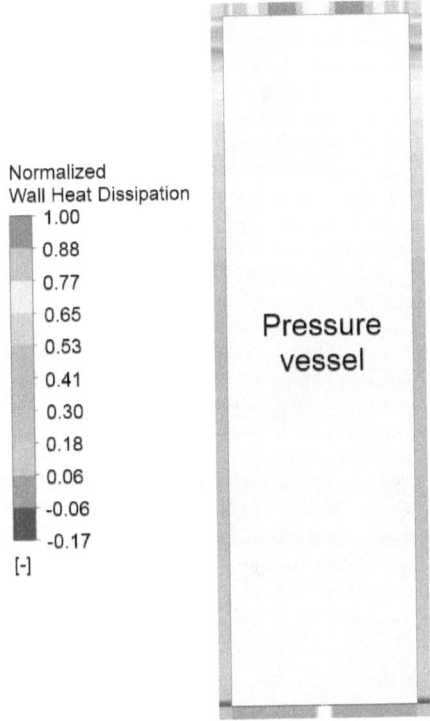

Figure 6.4. Heat flux profiles along the walls, the ceiling and the floor of the vessel. Positive values are defined as heat leaving the vessel.

By plotting the wall heat dissipation rate at the edges of the computational domain, shown in Figure 6.4, the areas with the highest amount of heat flux out through the vessel can be localized. As expected, a general increase of heat dissipation rate is seen towards the top as heat always strives to move upwards. A few local maxima can though be identified which are explained by a combination of high temperature and a high velocity, which result in a high enthalpy flow.

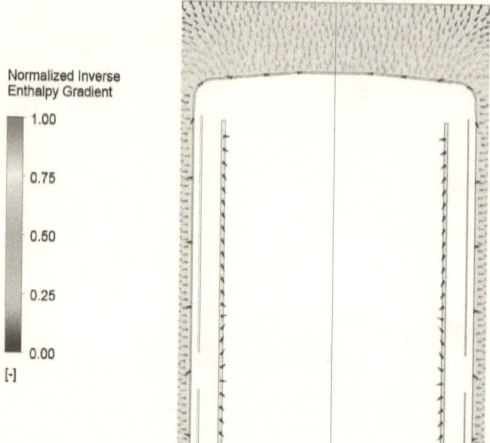

Figure 6.5. Upper part of the vessel with normalized vectors showing heat flux in the solid zones. The directions of the vectors are determined by temperature gradients and the vector colors are determined by enthalpy gradients.

An overview of heat flux in the solids is shown in Figure 6.5. The direction of the vectors illustrates the path that the conducted heat is taking and the color of the vectors determines the amount of energy that is being transported.

The highest energy transport per unit length is occurring through the sidewalls of the insulation mantle. Due to a thicker layer of insulation at the upper and lower part of the furnace, a higher thermal resistance is achieved which is restricting the upward and downward heat flux.

6.1.3 Validation

Measurement data is compared with the two solutions in the table below.

Table 6.2. Comparison between the true transient solution, the pseudo-transient solution (before mesh refinement) and measurement data.

Source	Total heat dissipation rate out through the vessel	Local temperature at upper part of vessel wall	Mass flow in cooling channels	Operating pressure
True transient solution	45450 W	125°C	1.62 kg/s	1311 bar
Pseudo-transient solution	45431 W	123°C	1.62 kg/s	-
Measurement data	29000-44000 W	125-150°C	-	1242-1253 bar

The two methods show close to identical results in all compared variables. The measurements may vary vastly from one HIP-cycle to another, so the measurement data that is collected from several logged HIP-cycles, is therefore presented in the form of ranges.

6.2 Cooling phase results

6.2.1 Transient solution

The temperature distribution during the first twenty seconds of the cooling phase simulation is visualized in the image sequence below.

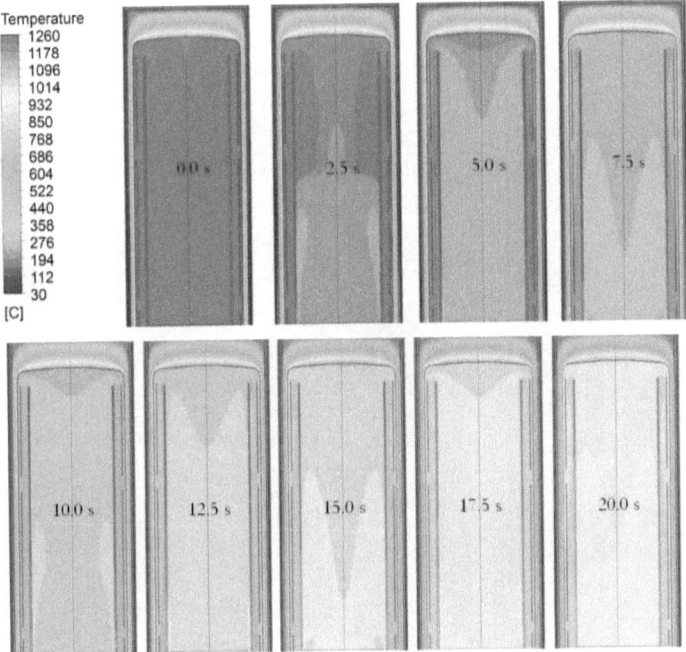

Figure 6.6. Overall temperature contours every 2.5 s during the first twenty seconds of the simulated cooling phase.

The temperature contours show how the loading section is cooled very uniformly and that a temperature decrease of approximately 900°C/min is achieved. Since the cooling process is working primarily by forced convection, the gas is cooled with a considerably higher rate than the solids since the cooling of the solids is highly dependent on their heat conducting abilities.

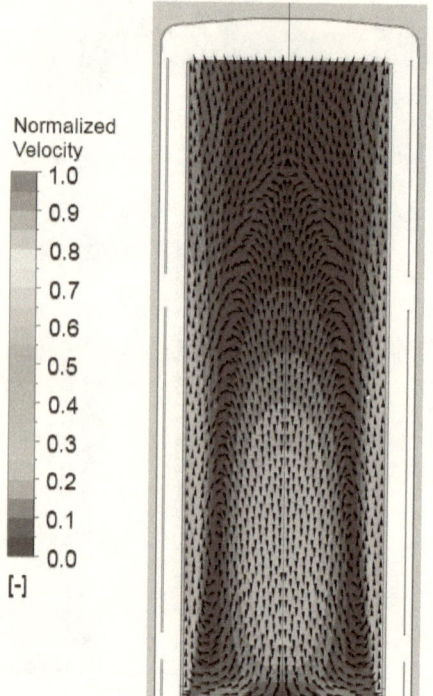

Figure 6.7. Velocity contour and velocity vectors in the loading section after 20 seconds of cooling.

In Figure 6.7 it is shown how the cold gas enters the furnace in the form of strong streams at the sides. This gives rise to additional gas movement and forming of vortices inside the furnace, which enhances the mixing of the gas, and thereby contributes to the uniformity of the cooling.

7 DISCUSSION AND CONCLUSIONS

The simulations of the sustained state and the transient phase were both successfully performed. Below follows discussions about the results and conclusions with the intention to answer the formulation of questions in the first chapter.

7.1 Discussion

7.1.1 Sustained state simulation

Considering the simulations of the sustained state, the true transient method required significantly more computational time than the pseudo-transient method, which was shown to be more time efficient than expected. The high simulation speed shown by the pseudo-transient approach is explained by the method's possibility to march in time with different time steps for the fluid zones and the solid zones. This becomes very efficient in these simulations, as the major bottleneck for the computational speed is to wait for the heat conduction to finish in the solids. These solids do however allow calculations with a higher time step than the fluid, as they do not involve convection of any property. This is taken advantage of with the pseudo-transient method.

Even though the pseudo-transient method converged in a much shorter time than the true transient method, the produced solutions show only minor differences between one another. The few differences that however do exist are mainly local varieties and appear in areas with relatively coarse grid sizes. The differences may therefore be linked to the mesh that eventually is too coarse in those areas. Another contributing factor to the comparison results could be that the true transient simulation has not fully reached steady state yet.

The gas flows and heat flux inside the HIP vessel are linked to the total heat dissipation rate out through the cylinder walls. A significant difference between the two solutions would therefore be possible to determine from the heat dissipation rates, which would then differ from one another.

Since the two solutions were overall close to identical and they showed a difference in dissipation rates of less than 0.1%, the two methods are thereby considered to be equally reliable. The pseudo-transient approach is however preferred because of its considerably lower computational time.

It should be noted that, during the evaluation of solving methods, various approaches that used regular steady state solvers were tested but were shown to not being able to successfully simulate the case without divergence problems.

In the sensitivity analysis, it became clear that it is of high importance to accurately resolve the strong temperature gradients in the mesh. Although they were originally resolved relatively well in this case, they could still be refined further, resulting in small impacts on the total heat dissipation rate of the vessel.

The validation of the simulations showed only minor differences in operating pressure and total heat dissipation rate compared to measurements. This however does not implicate that the simulations cannot be validated further. More validations of flow properties in key locations in the system are desired and would increase the reliability of the model. It should though be mentioned that a simulation, especially of this complexity, can never be fully consistent with reality as the reality will always have imperfections and variations of which all of them cannot be predicted. These kinds of unpredictable factors include imperfect manufacturing, equipment damages, daily differences in load structure and varying ambient conditions.

7.1.2 Cooling phase simulation

To be able to apply a fix velocity to a specific gas volume, it turned out that the pressure-velocity coupling had to be changed from the Coupled algorithm to a segregated algorithm, such as SIMPLEC. To avoid problems with divergence, the applied velocity furthermore had to be increased gradually from zero to the desired final value, in the time span of a few milliseconds.

The transient solution showed a furnace cooling rate of 900°C/min, which is approximately a three times higher rate than what has been observed in measurements. This was however expected, as the simulated vessel contains no load and the measured cases involve loads of at least 100 kg, which significantly increases the total thermal mass of the furnace, and thereby also increase the required amount of cooling.

Due to the high level of mixing in the loading section, the incoming cold gas successfully managed to cool the furnace in a very uniform approach without creating any strong local temperature differences. It is however possible that the cooling performance will behave differently when considering future cases that include loads.

The transient cooling simulation was found to be a very time-consuming task that produced a 20 seconds long cooling simulation in approximately two weeks. Future more complex models can be assumed to be even more time-consuming. However, suggestions for decreasing the amount of required computational time in future simulations are presented in chapter 8.

7.2 Conclusions

Two simulation models have successfully been developed for analyzing the gas flow and the heat conduction that is taking place inside the HIP apparatus. The first simulation model yields a steady state solution of the sustained state using a pseudo-transient approach. The second simulation model yields a transient solution for the cooling phase of the HIP-cycle using a regular transient solver.

The steady state solution has been validated through pressure measurements, measured heat dissipation rate from the vessel wall, as well as one local temperature measurement close to the vessel wall.

The transient solution has not been validated since the simulated cooling rate cannot be compared to measurement data. The reason is that the logged HIP-cycles are involving loads while the simulation does not.

The aim of this work has been to introduce the use of CFD to the discipline of hot isostatic pressing. As the technology, to the author's knowledge, has not previously been used in this field, no guarantee has existed in advance that simulations would be possible. Neither has any modeling guidelines been available during the simulation process. With the achieved results, CFD has now been introduced into the HIP field and the presented modeling methods may serve as guidelines for future simulations.

8 RECOMMENDATIONS AND FUTURE WORK

Even though the performed simulations successfully produced satisfactory results, the improvement of the models should nevertheless continue to further increase the accuracy and reliability of future simulations. Below follows some guidelines for what the next steps in this development should include.

8.1 Recommendations

For future steady state simulations that are similar to the one performed in this work, a pseudo-transient method is preferred, as it will save a significant amount of time while producing results as reliable as a true transient method.

In the performed simulations, the insulation zones have been modeled as true solids, which is a simplification that is inconsistent with reality. A more accurately modeled insulation mantle should take account for the porosity of the insulation material so that the gas can move around inside the insulation zones. This would enable the capturing of convection phenomena that are working inside the porous material, resulting in a more correctly modeled heat transport in the insulation zones. It should be noted however that a resulting increase in computational time is to be expected.

An increased mesh resolution in the cooling phase simulation could result in that some additional flow phenomena are captured. To avoid an undesired increase in mesh size, dynamic mesh adaption should be considered. It can be used to dynamically refine the mesh at certain time intervals at temperature gradients and velocity gradients, as they develop and relocate in the domain.

When implementing loads in future simulations, giving the ability to investigate the cooling rate of the load, it should be noted that the mesh size and overall complexity of the problem will increase. It is therefore suggested that the load is simplified in one way or another. One approach is for instance to have a cylinder-shaped load that is covering the complete load section. The cylinder may then consist of a porous material that retards the flow to an appropriate degree and may also have a thermal mass corresponding to the one of the actual load.

When increasing the complexity of future simulations models, such as by implementing loads, it is recommended to reduce the problem by delimiting the computational domain if possible, or split it into several subdomains so that they can be simulated separately. A logical splitting of the computational domain is obtained by separating the hot furnace from the cooling channels so that the two areas are instead being considered as two isolated flows. The option of dividing the problem should be considered for both steady state simulations and transient simulations that are performed in the future.

If the outcome of a future simulation has an important purpose such as evaluating design alterations, it is recommended that a stronger validation of the model should be performed to increase its reliability. To achieve a strong validation it is recommended to compare the solution with several local temperature measurements in a number of key locations.

8.2 Future work

One of the next steps for the future work is to progress from the 2D-axisymmetric model to a full-scale 3D analysis. This would open up possibilities to correctly account for factors such as periodic and non-periodic details in the geometry, eddies in three dimensions and swirling flow phenomena.

The k-ω turbulence model was chosen for the performed simulations due to its versatility and simplicity. No definite evaluation of different turbulence models has however been conducted in this work but should be carried out in the future as it may have an impact on the solutions. The possibility of rotating flows in the furnace should not be excluded if performing a future 3D-analysis, in which case the Reynolds stresses model could be used to efficiently resolve the turbulence.

Additional simulations focusing on individual parts in the HIP vessel such as the cooling channels and the ejector is also of high interest. The computational domain would in such simulations be delimited to only enclose the investigated component and the nearby surroundings. The boundary conditions could be determined from an overall simulation of the complete vessel such as the ones presented in this work.

Fluid-structure interaction (FSI) deals with the interaction between a fluid and a structure on which the fluid is exerting a pressure. The fluid may cause the structure to deform, which in turn will affect the fluid flow itself. The technology could for instance be used in optimization of the cooling system by determining if a certain design alteration or a certain velocity increase will result in deformations of the structure.

REFERENCES

Zimmerman, F. X., & Toops, J. (2008). *Hot Isostatic Pressing: Today and Tomorrow*. Avure Technologies Inc.

Welding Technology Institute of Australia. (2006). Hot Isostatic Pressing (HIP) for Manufacture of Orthopaedic Implants. Silverwater. Retrieved from WTIA.

Yang, Y., de Jong, R. A., & Reuter, M. A. (2005). *Use of CFD to predict the performance of a heat treatment furnace*. Trondheim: 8th International Conference on COMPUTATIONAL FLUID DYNAMICS in the Oil & Gas, Metallurgical and Process Industries.

Vijayan, P. K., & Nayak, A. K. (2010). *Experimental validation and data base of simple loop facilities*. India.

ANSYS Inc. *Documentation for FLUENT 13.0*.

ASEA Metallurgy. (1984). *Quintus Isostatic Pressing Technology*. Västerås.

Bakker, A. (2002). *Applied Computational Fluid Dynamics: Lecture 5 - Solution Methods*. Retrieved May 11, 2012, from The Colorful Fluid Mixing Gallery: http://www.bakker.org/dartmouth06/engs150/05-solv.pdf

Peric, M., & Bertram, V. (2011). Trends in Industry Applications of CFD for Maritime Flows. Berlin: 10th International Conference on Computer and IT Applications in the Maritime.

Sayma, A. (2009). *Computational Fluid Dynamics*. BookBoon.

Rao, S. P. (2010). Modeling of Turbulent Flows and Boundary Layer. In H. O. Woo, *Computational Fluid Dynamics*. Vukovar: Intech.

APPENDIX A: SOLUTION COMPARISON FOR SUSTAINED STATE

The true transient method and the pseudo-transient method are compared by analyzing the differences between the two resulting solutions. The contour plots below show the resulting differences after subtracting the pseudo-transient solution from the transient solution.

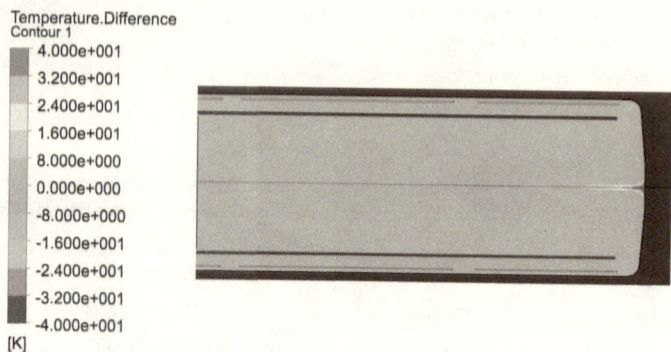

Resulting temperature field after subtracting the pseudo-transient solution from the transient solution. Generally a temperature difference smaller than 3 K in the green areas in the figure.

Resulting density field after subtracting the pseudo-transient solution from the true transient solution. Generally a density difference smaller than 3 kg/m3 in the green areas in the figure.

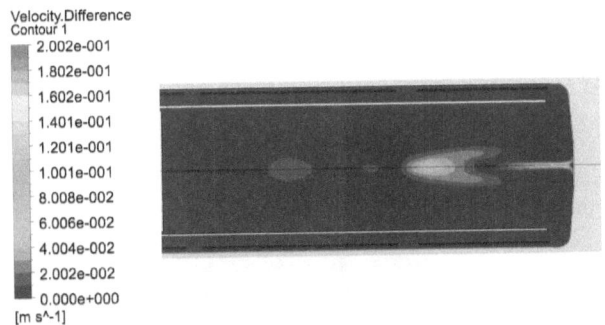

Resulting velocity field after subtracting the pseudo-transient solution from the true transient solution.

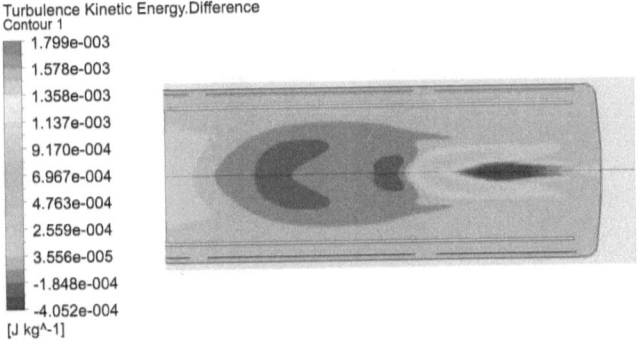

Resulting TKE field after subtracting the pseudo-transient solution from the true transient solution. Generally a difference in turbulence kinetic energy below 0.001 J/kg in the blue areas in the figure.

APPENDIX B: TKE PLOT OF SUSTAINED STATE

The figure below show a contour of TKE that is overlaid with velocity vectors.

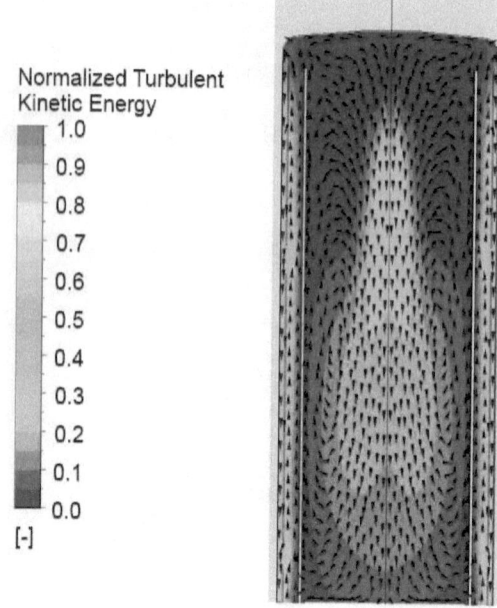

APPENDIX C: RESULTS FROM SENSITIVITY ANALYSIS

The impact of the sensitivity analysis is determined by analyzing the differences between the solution before and after the mesh refinement. The contour plots below show the differences in several flow properties after subtracting the refined-mesh-solution from the original-mesh-solution.

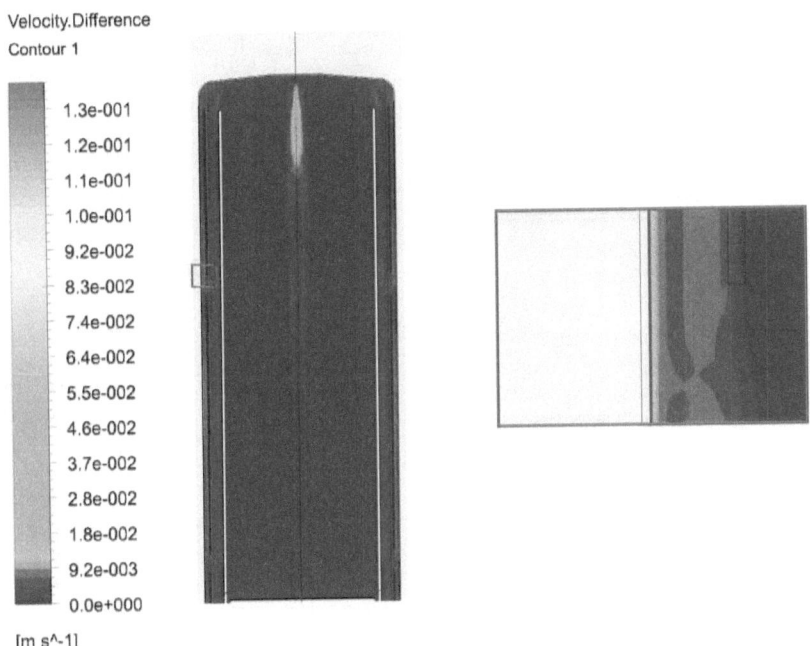

Contour plots showing the resulting velocity differences obtained from subtracting the refined-mesh-solution from the original-mesh-solution. To the left are the velocity differences in the furnace and to the right are a detailed close-up.

Further refinement did not have any impact on the solution.

I want morebooks!

Buy your books fast and straightforward online - at one of the world's fastest growing online book stores! Environmentally sound due to Print-on-Demand technologies.

Buy your books online at
www.get-morebooks.com

Kaufen Sie Ihre Bücher schnell und unkompliziert online – auf einer der am schnellsten wachsenden Buchhandelsplattformen weltweit! Dank Print-On-Demand umwelt- und ressourcenschonend produziert.

Bücher schneller online kaufen
www.morebooks.de

OmniScriptum Marketing DEU GmbH
Heinrich-Böcking-Str. 6-8
D - 66121 Saarbrücken

Telefax: +49 681 93 81 567-9

info@omniscriptum.de
www.omniscriptum.de